Algae- A Survey from Temples

Darani Vasudevan

Preface to the book

This book is an account on the species of algae colonizing the wall, floor, tank surface and stone sculptures of three different temples of Erode, a district of Tamil Nadu, India. I have carried out this work as a research and the dissertation was submitted to the Periyar University, Salem, Tamil Nadu (India). Hope the book will be a reference to those who like to carry out research in the field of Phycology (Algae).

-V.Darani M.Sc., M.Phil., (SET)

Contents inside....

Introduction

Materials used and Methodology

Results

Discussion

Summary and conclusion

Bibliography

Chapter 1: Introduction

Algae are the important components of the biosphere. They exist in many forms ranging from simple unicellular species to complex multicellular species. The blue green algae (Cyanophyceae) were the primitive photosynthetic organisms to evolve on Earth. Blue green algae along with bacteria are responsible for the composition of oxygen in the primitive atmosphere. Algae lay the foundation for the aquatic food chains and constitute 70% of oxygen we breathe. The algal colonization depends primarily on the nutrition availability. Biofilms are the thin mucilaginous layers of varied colours formed by the growth of algae and cyanobacteria on the exposed surface of buildings during rainy season. Thus biofilms get slowly dried after rain and remain perched throughout the year to the stones (Karande *et al*, 2012). These terrestrial algae exist as epiphytes on tree barks, as epiliths on rocks and surfaces of buildings. They prefer moist and wet places for growth and thrive well in humid conditions. Algae, heterotrophic bacteria and fungi are the main colonizers of the biofilms. The phototrophic algae and blue green algae have been considered to be the primary colonizers, conditioning the inert surfaces for the growth of heterotrophic organisms, such as fungi (Crispim *et al.,* 2002).

Temples are regarded as an important part of World's cultural heritage. Naturally the stones undergo the process of weathering through which they are deteriorated. The weathering is carried out by physical, chemical or biological processes. One of the most complex problems in the conservation buildings and monuments is to contrast biodeterioration (Balu Bhavani *et al*., 2013).

Environmental conditions and physicochemical properties of a material determine the growth and colonization of algae and other microbes. Algal colonization results in the staining of the surface by biogenic pigments and also by the production of extracellular polymeric substances (EPS). They alter the pore size, moisture circulation pattern and temperature responses. The deposition of surfactants by the microorganisms alters the water permeability of the substances. They enhance the accumulation of pollutants in the affected area. The activity of microbes in the alteration and deterioration of stone and concrete walls has been frequently neglected. There is a need for documentation of species colonizing the stone and concrete structures of the temples and other monuments for identifying the conditions favouring such growth. Formulation of a remedy will not be possible unless or until the identification of the cause of the deterioration is done.

The present work aims to document the algal flora from the temple wall, temple tank, temple floor and temple statue of three different temples of Erode district, Tamil Nadu.

Chapter 2: Materials used and methodology

Algal sample collection

The algal samples were collected from three temples of Erode district namely Sree Cheliyandiamman temple, Bhavani, Sree Seedeviamman temple, Kanchikovil and Arulmigu Arudra Kabaleeshwarar temple, Erode in the month of December, 2015.

Sree Cheliyandiamman temple is located at 11°27'N 77 °41'E and situated in Chinapuliyur village which is about 18 Km towards North from district headquarters, Erode. The temple is at an altitude of 175 m above sea level. Sree Seedeviamman temple is located at 11.36944 °N 77.59667 °E and situated in Kanchikovil, Nasiyanur, Erode district. The temple is at an altitude of 292 m bove the sea level. Arulmigu Arudra Kabaleeshwarar temple is located at 11 °22'N 77 °44'E and situated in Erode district. The temple is at an altitude of 183 m above sea level.

Algal flora Sampling

Algal samples were collected at three sites namely temple wall, temple floor and temple water tank in Sree Cheliyandiamman temple and Sree Seedeviamman temple respectively. In case of Arudra Kabaleeshwarar temple the sample sites were temple statue, temple floor and temple wall. These sites were chosen based on the visual appearance of algal growth on temple areas. All sampled sites were moist to dry. The samples from each area were collected in sterile polythene bags using scalpel and pliers. No original stone material was damaged during collection. Algal samples were collected from the prefixed by the serial number code of the habitat and district code. Accordingly,

- TN/10/SA(1)/WL
- TN/10/SA(2)/FR
- TN/10/SA(3)/TK

- TN/10/CA(4)/WL
- TN/10/CA(5)/FR
- TN/10/CA(6)/TK
- TN/10/KA(7)/ST
- TN/10/KA(8)/FR
- TN/10/KA(9)/WL

The following were abbreviations of the above said algal sample samples, site collected

TN- Tamil Nadu

10- Erode district code

SA- Sree Seedeviamman temple

CA-Cheliyandiamman temple

KA-Arulmigu Arudra Kabalishwarar temple

1-9- Sample numbers

WL- Temple wall

FR- Temple floor

TK- Temple tank

ST- Temple statue

For example TN/10/SA(1)/WL indicates that the first sample was from the wall Sree Seedeviamman temple of Erode district, Tamil Nadu. In the same way TN/10/SA (2)/ FR indicates that the second sample was from floor of Sree Seedeviamman temple of Erode district, Tamil Nadu and so on.

Collection methods

The epilithic floras were collected by using two sampling methods,

Method I

A soft tooth brush was used to lightly scrub in a circular motion on a small area on temple monuments and then rinsed with distilled water. The tooth brush was cleaned with distilled water after each sampling. This rinse water was collected in a sterile polythene bag and stored in the laboratory for further analysis.

Method II

The top layer of sediment with wall flora was removed with a single edge razor blade, placed in a sterile polythene bag and stored in the laboratory for further analysis.

Culturing of samples

Algal samples were collected from the stations mentioned earlier and were cultured using the Bold's Basal medium.

Preparation of Bold's Basal medium

BBM is an algal medium that has ben used to grow a variety of green algal cultures. Media may be used as liquid or solidified by agar. This medium is highly enriched. The composition of BBM is given in the Table 1.

Table 1: Liquid Bold's Basal Medium

MACRONUTRIENTS

Sodium nitrate (NaNO$_3$)	10g
Calcium chloride dehydrate (CaCl$_2$.2H$_2$O)	1g
Magnesium sulphate heptahydrate (MgSO$_4$.7H$_2$O)	3g
potassium hydrogen orthophosphate (K$_2$HPO$_4$)	3g
Potassium dihydrogen phosphate (KH$_2$PO$_4$)	7g
sodium chloride (NaCl)	1g

Dissolved in 400ml distilled water

MICRONUTRIENTS

Ethylene Diamine Tetra Acetic acid (EDTA)	50g
Potassium hydroxide (KOH)	31g

Dissolved in 1 litre distilled water

Ferrous sulphate heptahydrate (FeSO$_4$.7H$_2$O)	4.98g
Sulphuric acid (H$_2$SO$_4$)	1ml

Dissolved in 1 litre distilled water

Boric acid (H$_3$BO$_3$)	11.42g
Zinc sulphate heptahydrate (ZnSO$_4$.7H$_2$O)	8.82 g
ManganousII chloride tetrahydrate	1.44g

(MnCl$_2$.4H$_2$O)	
Molybdenum (MoO$_3$)	0.71g
Cupric sulphate pentahydrate (CuSO$_4$.5H$_2$O)	1.57g
Cobalt (CO(NO$_3$)$_2$.6H$_2$O)	0.49g

Dissolved in 1 litre distilled water.

10 ml macronutrients + 1 ml each micronutrients +936 ml deionized water altered to pH 7 by adding 0.1N NaOH.

The medium was sterilized in a pressure cooker for 15-20 minutes. After sterilization the medium was allowed to cool.

Because of some practical difficulties in identifying algae, the collected samples were cultured in liquid Bold's Basal medium. Algae were collected and brought to laboratory and inoculated in culture plates containing liquid Bold's basal medium and placed in a culture rack under a cool white fluorescent tube light at photon flux densities of about 20E/M-2/8-1 for about 15 days.

Documentation

Preparation of slides

Semi permanent slides were made for each sample. Depending on the abundance of the algal population 2 to 5 replicates were prepared. The samples were taken in a petridish containing water and then spreaded in it. From that thin portion or drop of water were placed on a clean microslide (Bluestar). The sample is covered with coverslip (Bluestar) after adding glycerine to it. While using coverslip care should be taken to avoid formation of air bubbles. The edges of the coverslip and the surface of the slide around the coverslip were sealed by using neutral nail polish or DPX mountant.

Screening of slides for Algae

The slides were observed under low and high power pre calibrated Olympus microscope using natural daylight. Observations for each sample were recorded on separate sheet. Each sample were analysed. In the case of algae whose identity could be recognized at once, names were noted down tentatively. For all algae rough drawings were made and ocular measurements for the cell dimensios were recorded, which could be converted into microns using the calibration value. After screening all the slides photomicrographs were taken using Nikon camera fitted to the eye piece of binocular microscope.

Identification of Algae

The algae mounted in the slides and photographs were identified using the standard works of Anand (1980) and Gonzalves and Kamal (1958).

Percentage of frequency (Anisha and Kulathuran, 2010)

Frequency occurrence was calculated as follows,

$$\% \text{ frequency} = \frac{\text{Number of samples in which specific organisms occurred}}{\text{Total number of samples examined.}}$$

Based on the frequency occurrence the algae were grouped as rare (0-25% frequency), occasional (26-50 % frequency), Frequent (51-75% frequency), Common (76-100 % frequency) species.

Chapter 3: Results

An algal survey was carried out in three temples of Erode district namely, Sree Cheliyandiamman temple, Bhavani, Sree Seedeviamman temple, Kanchikovil and Arulmigu Arudra Kabaleeshwarar temple, Erode. The algal collection was carried out in three sites temple floor, temple wall and temple tank and temple statue. A total of 49 species belonging to 18 genera were recorded. These genera belonged to three classes namely, cyanophyseae, chlorophyceae and bacillariophyceae. Of the total 49 species, 18 species were recorded from wall, 18 species were recorded from floor, 19 species were recorded from tank and 10 species were recorded from statue.

Algal flora of Sree Cheliyandiamman temple

A total of 36 species belonging to 18 genera were recorded from the temple. These genera belonged to three classes namely cyanophyceae, chlorophyceae and bacillariophyceae. Of the total 36 species, 14 species were recorded from wall, 11 species were recorded from floor and 11 species were recorded from tank.The species included 6 colonial forms, 4 unicellular forms, 3 diatoms and 23 filamentous forms. The species of *Aphanocapsa, Synechocystis, Lyngbya, Oscillatoria, Phromidium, Anabaena* and *Nostoc* were found in wall, the species of *Aphanocapsa, Gloeocapsa, Synechococcus, Lyngbya, Oscillatoria* and *Phromidium* were found in floor. The species of *Chroococcus, Microcystis, Merismopedia, Arthrospira, Plectonema, Bracteacoccus, Rhizoselenia* and *Navicula* were found in tank. The colonial forms belonged to genera *Aphanocapsa, Chroococcus, Microcystis, Rhizoselenia* and *Navicula*. The unicellular forms belonged to genera *Synechocystis, Bracteococcus, Synechococcus* and *Gloeocapsa*. The filamentous forms belonged to genera *Lyngbya, Oscillatoria, Plectonema, Phromidium, Anabaena* and *Nostoc*. The dominant genera were *Lyngbya* and *Oscillatoria* consisting of 7 species each followed by *Phromidium* which included 4 species followed by *Plectonema, Navicula, Chroococcus, Aphanocapsa* consists of 2 species each and the least occurring genera were *Rhizoselenia, Bracteacoccus, Anabaena, Nostoc, Synechococcus, Synechocystis, Arthrospira, Gloeocapsa, Microcystis* and *Merismopedia* with 1 species each. The survey recorded 3 diatoms namely, *Rhizoselenia sp., Navicula bacillum* and *Navicula laterostrata*.

Algal flora of Sree Seedeviamman temple

A total of 38 species belonging to 15 genera were recorded from the temple. These genera belonged to three classes namely cyanophyceae, chlorophyceae and bacillariophyceae. Of the total 38 species, 12 species were recorded from wall, 14 species were recorded from floor and 12 species were recorded from tank. The species included 7 colonial forms, 4 unicellular forms, 2 diatoms and 25 filamentous forms. The species of *Aphanocapsa, Chroococcus, Lyngbya, Oscillatoria, Phromidium* and *Nostoc* were found in wall, the species of *Chroococcus, Gloeocapsa, Microcystis, Synechoccus, Lyngbya, Oscillatoria* and *Phromidium* were found in floor. The species of *Aphanocapsa, Chroococcus, Gloeocapsa, Merismopedia, Lyngbya, Phromidium, Anabaena, Plectonema, Bracteacoccus, Pinnularia* and *Navicula* were found in tank. The colonial forms belonged to genera *Aphanocapsa, Chroococcus, Microcystis, Rhizoselenia, Pinnularia* and *Navicula*. The unicellular forms belonged to genera *Synechococcus, Bracteococcus* and *Gloeocapsa*. The filamentous forms belonged to genera *Lyngbya, Oscillatoria, Plectonema, Phromidium, Anabaena* and *Nostoc*. The dominant genera was *Lyngbya* consisting of 10 species followed by *Oscillatoria* which included 7 species followed by *Phromidium* which included 4 species, *Chroococcus* with 3 species followed by *Plectonema, Gloeocapsa and Aphanocapsa* consists of 2 species each and the least occurring genera were *Microcystis, Merismopedia, Synechococcus, Anabaena, Nostoc, Bracteacoccus, Navicula* and *Pinnularia* with 1 species each. The survey recorded 2 diatoms namely, *Navicula bacillum* and *Pinnularia interrupta*.

Algal flora of Arulmigu Arudra Kabalishwarar temple

A total of 29 species belonging to 12 genera were recorded from the temple. These genera belonged to class cyanophyceae. Of the total 29 species, 11 species were recorded from wall, 8 species were recorded from floor and 10 species were recorded from statue. The species included 6 colonial forms, 3 unicellular forms and 20 filamentous forms. The species of *Aphanocapsa, Synechocystis, Arthrospira, Lyngbya, Oscillatoria, Phromidium, Anabaena* and *Nostoc* were found in wall, the species of *Microcystis, Synechococcus, Lyngbya, Oscillatoria* and *Phromidium* were found in floor. The species of *Aphanocapsa, Chroococcus, Glepcapsa, Microcystis, Lyngbya, Oscillatoria* and *Phromidium* were found in statue. The colonial forms belonged to genera *Aphanocapsa, Chroococcus* and *Microcystis*. The unicellular forms belonged to genera *Synechocystis, Synechococcus* and *Gloeocapsa* The filamentous forms belonged to genera *Lyngbya, Oscillatoria, Phromidium, Anabaena* and *Nostoc*. The dominant genera was *Lyngbya* consisting of 8 species followed by *Oscillatoria* which included 5 species followed by *Phromidium* consisting of 4 species, *Microcystis* consisting of 3 species, followed by *Aphanocapsa* with 2 species and the least occurring genera were *Chroococcus, Gloeocapsa, Synechococcus, Synechocystis, Arthrospira, Anabaena* and *Nostoc* with 1 species each.

Among surveyed algae cyanophyceae was dominant with 89.79% of the total species and 77.77% of the total genera being represented by 44 species and 14 genera. The class bacillariophyceae was the second highest in the order of dominance represented by 4 species belonging to 3 genera and they accounted 8.16% and 16.66% of the total species and genera respectively. Class chlorophyceae occupied the third place in the order of dominance being represented by 1 species and 1 genera and accounted by 2.04% and 5.55% of the total species and genera respectively.

Frequency of algal species (Anisha and Kulathuran, 2010)

Frequency occurrence was calculated as follows,

% frequency = $\dfrac{\text{Number of samples in which specific organisms occurred}}{\text{Total number of samples examined.}}$

Based on the frequency occurrence the algae were grouped as rare (0-25% frequency), occasional (26-50 % frequency), Frequent (51-75% frequency), Common (76-100 % frequency) species.

Calculation of percentage frequency of the noted species revealed that species like *Aphanocapsa grevelli, Microcystis aeruginosa, Synechococcus elongates, Lyngbya epiphytica, Oscillatoria acuminate, Phormidium jadinianum, Anabaena fertilissima, Nostoc calcicola, Plectonema nostocorum* etc., are commonly present in all the three surveyed temples with 100% occurrence. Species like *Rhizoselenia sp. ,Navicula bacillum, Oscillatoria curviceps, Phormidium tenue, Lyngbya semiplena, Chroococcus minutes, Synechocystis pevalekii* etc., are occasionally present with 50% frequency. These are encountered in two of the surveyed three temples. *Chroococcus minor, Chroococcus turgidus, Lyngbya subtilis, Phormidium mucosum, Pinnularia interrupta* etc., are the rarely occurring species present in only one place and are with 25% frequency.

Taxonomic account

Taxonomic account of the algal species recorded is given in the following pages alphabetically in the order of Cyanophyceae, Chlorophyceae and Bacillariophyceae. Cell dimensions and important characteristics are presented for each algal species. This is followed by a note on their general habitat. Wherever some interesting observations made it is indicated as a note. Binomials used are as per the manuals of research optical referred to (cited in materials and methods). The latest nomenclatural status of the taxon is not taken into account.

Cyanophyceae

Chroococcaceae

1. *Aphanocapsa grevillei* (Hass.)Rabenh

Colony in the form of a gelatinous mass, thallus attached, cells upto 5.6 μm, cells submerged, individual envelope not distinct.

2. *Aphanocapsa pulchara* (Kutz.)Rabenh

Thallus planktonic, cells upto 4.5 μm diameter.

3. *Chroococcus minor* (Kutzing)Nageli

Thallus made of cells in groups. Sheath not lamellated, cells narrower, cells upto 4.0 μm broad.

4. *Chroococcus indicus* Bernard

Thallus large gelatinous colonies, colonies attached, cells upto 11.0 μm broad.

5. *Chroococcus turgidus* (Kutzing)Nageli

Thallus made of cells in groups, sheath distinct lamellated, 18.0 μm broad.

6. *Chroococcus minutes* (Kutzing)Nageli

Thallus made of cells in groups, sheath not lamellated, cells narrower upto 7.0 μm broad.

7. *Gloeocapsa nigrescens* Nag

Thallus mucilaginous, sheath unlamellated, thallus blue green, cells upto 6.8 μm broad

8. *Gloeocapsa atrata* (Turp.)Kutz

Thallus mucilaginous, sheath unlamellated, thallus blackish, cells 4.5 µm broad.

9. *Merismopedia punctata* Meyen

It is ovoid or spherical in shape and arranged in rows and flat, forming rectangular colonies held together by a mucilaginous matrix. Species of this genus divide in only two directions creating a characteristic grid like pattern.

10. *Microcystis flos-aquae* (Wittr.)Kirchner

Colonies not clatharate, cells arranged closely, margin of colonial mucilage diffluent, cells 6.5 µm broad.

11. *Microcystis aeruginosa* Kutzing

Colonies clathrate, cells 6.5 µm broad.

12. *Microcystis robusta* (Clark)Nygarrd

Colonies irregularly overlapping net like, cells 10-12 µm diameter. Cells in homogenous, colourless, diffluent mucilage densely packed, colonies not clathrate. Cells arranged closely. Gas vacuoles commonly seen.

13. *Synechococcus elongatus* Nag

Cells oblong, ellipsoidal or cylindrical, generally single without any mucilage envelope. The cells were 1.9- 2.6 µm broad and 3.6 µm long.

14. *Synechocystis pevalekii* Ercegovi

Cells free spherical single or 2 together after division, without mucilaginous cells rounded, 2.5-3.5 µm broad.

Oscillatoriaceae

15. *Arthrospira platensis* Gomont

Spirals away from one another, more than 15 µm broad. Trichomes 7.8µm broad, unconstricted.

16. *Lyngbya kuetzinghiana* kirchner

Filaments entangling forming expanded thallus, sheath thin and diffluent, filaments straight, trichomes narrower, trichomes 4.5 µm broad.

17. *Lyngbya subtilis* (Kutzing) Hansging

Filaments free, sheath thin and diffluent, filaments narrower, filaments 2.5 µm broad.

18. *Lyngbya aerugino-coerulae* Agardh ex. Gomont

Filaments free, sheath thin and diffluent, filaments 9.5 µm broad.

19. *Lyngbya putealis* Mont. ex. Gomont

Filaments entangling forming expanded thallus, sheath thin and diffluent, filaments straight, trichomes constricted, trichomes 7.5 µm broad.

20. *Lyngbya diguetii* Gomont

Colony a mat of filaments, 2 mm thick, filaments 2.5-3 µm wide, entangled towards base, but ends more or less straight towards apex. Cells 2-2.8 µm wide, 1-3.7 µm long, quadrate or shorter than wide, cross wall is not narrowed, end cells rounded, without calyptras. Sheath thin, colourless.

21. *Lyngbya epiphytica* Hieron

Filaments epiphytic on other algae, firmly attached and often coiled round the host. Filaments 1.5-2 µm wide. Cells 1- 1.5 µm wide, 1-2 µm long. Cells slightly shorter or longer than wide. Cross walls not narrowed, end cells rounded, not tapering, sheath thin, colourless.

22. *Lyngbya semiplena* (G.Ag.)J.Ag.ex.Gomont

Filaments entangling forming expanded thallus. Sheath thick, end cells capitates, filaments 16 μm broad.

23. *Lyngbya majuscula* Harvey ex. Gomont

Filaments entangling forming expanded thallus, sheath thick and cells rounded, filaments narrower, 9 μm broad.

24. *Lyngbya major* Menegh.ex.Gomont

Filaments entangling forming expanded thallus, sheath thick, end cells flat with outer membrane, filaments upto 15 μm broad.

25. *Lyngbya confervoides* L.Ag.ex.Gomont

Filaments entangling forming expanded thallus, sheath thick, end cells rounded, filaments narrower, filaments 20 μm broad.

26. *Lyngbya dendrobia* Brubl.ed.Biswas

Filaments entangling forming expanded thallus. Sheath thick. End cells rounded. Filaments narrower. Filaments upto 9 μm broad. Filaments narrower.

27. *Oscillatoria acuminata* Gomont

Trichomes not constricted at cross walls. Trichomes attenuated. Tip of trichome bent. Cells longer than broad. Trichome straight slightly bent 4 μm broad.

28. *Oscillatoria princeps* Vaucher ex. Gomont

Trichomes not constricted at cross walls. Trichomes attenuated. Tip of trichome not bent. Cells broader. End cells convex with thick outer membrane, trichomes 13 μm broad.

29. *Oscillatoria limosa* Ag ex. Gomont

Trichomes not constricted at cross walls. Trichomes not attenuated. Cells broader than long, end cells with thick outer membrane. Trichome 11 µm broad.

30. *Oscillatoria pseudogeminata* Schmid

Trichomes not constricted at cross walls, trichomes not attenuated. Cells as long as or longer than broad. Cells quadratic upto 3 µm broad.

31. *Oscillatoria willei* Gardner ex. Drouet

Trichomes not constricted at cross walls. Trichomes not attenuated. Cells longer than broad, granules absent.

32. *Oscillatoria subbrevis* Schmid

Trichomes not constricted at cross walls. Trichomes not attenuated. Cells broader than long. End cells rounded. Trichomes 5.5 µm broad.

33. *Oscillatoria amoena* Gomont

Trichomes not constricted at cross walls. Trichomes attenuated. Tip of trichome not bent. Cells broader. End cells convex with thick outer membrane, trichomes 13 µm broad.

34. *Oscillatoria curviceps* Agardh ex. Gomont

Trichomes not constricted at cross walls, trichomes not attenuated, cells broader tham long, end cells rounded, trichomes 15.0 µm broad.

35. *Phormidium tenue* (Menegh.)Gomont

Trichomes constricted at cross walls, trichomes attenuated. End cells acute conical. Filaments upto 2.5 µm broad.

36. *Phormidium fragile* (Menegh.)Gomont

Trichomes constricted at cross walls. Trichomes attenuated. End cells acute conical. Filaments upto 2.5 µm broad.

37. *Phormidium mucosum* Gardner

Trichomes constricted at cross walls, trichomes not attenuated, trichomes flexuous, 7.0 µm broad.

38. *Phormidium retzei* Kutzing ex. Gomont

Trichomes unconstructed, sheath thin, diffluent, trichomes not attenuated, end cells not capitates. End cells rounded with outer thickened membrane.

39. *Phormidium jadinianum* Gomont

Trichomes constricted at cross walls, trichomes attenuated, end cells conical, filaments parallel, 5.5 µm broad.

40. *Phormidium ambigum* Gomont

Trichomes constricted at cross wall, trichomes not attenuated, trichomes flexous, trichomes 7 µm broad.

Nostocaceae

41. *Anabaena fertilissima* Prasad

Akinetes away from the heterocyst, akinetes spherical, trichomes 5 µm broad. Akinetes in long chains, 7 µm broad.

42. *Nostoc calcicola* ex. Born

Thallus without firm outer layer, soft formless, trichomes less densely coiled. Thallus macroscopic. Trichomes not densely coiled, spores varying, cells barrel shaped, shorter or longer than broad. Trichomes upto 4 µm broad. Spores spherical, 4.5- 5.8 µm diameter.

Scytonemataceae

43. *Plectonema nostocorum* Bornet ex. Gomont

Filaments usually reported living inside mucilage of old colonies of other algae, straight or bent, branching sparse, usually single cells cylindrical, 0.7-1.5 µm wide, 2-3 µm long, end cells rounded, sheath thin and colourless.

44. *Plectonema radiosum* (Schiederm.)Gomont

Filaments irregularly curved, rounded or spherical, dull green, richly false branching, sheath in lower part of the filament. Cells distinctly constricted at the cross walls, 11-20 µm broad, 3-10 µm long, blue green, end cells rounded.

Chlorophyceae

45. *Bracteacoccus minor*(Chodat petrova)

Cells spherical, single or in group, globose, polygonal, parietal chloroplasts many without pyrenoid. 6.8 to 23 µm diameter.

Bacillariophyceae

46. *Rhizoselenia* sp.Brightwell

Cells cylindrical 77 µm in length and 8-9 µm in breadth. Intercalary bands arranged in several pervalvar series, scale like, rhomboidal to almost square in outling at the centre of the cells. Sides slightly wavy, calyptras flat with very clear impression of the sister valve.

47. *Navicula bacillum* Ehrenberg

Valves 30 µm long and 7 µm broad, linear with straight or slightly convex sides and broadly rounded ends, raphe thin and straight, axial area large, rounded.

48. *Navicula laterostrata* Hustedt

Valves elliptic lanceolate with broadly rounded and capitates ends. Axial area very narrow, central area big, rounded. Length- 20-25 µm ,breadth- 6-9 µm

49. *Pinnularia interrupta* Smith

Sides of the valve straight and parallel. Striations interrupted at the centre of the valve. Ends not very markedly capitates. Length-225-29 µm, breadth 4-6 µm

Table 2- Algal Flora from temples their cell nature, percentage frequency, frequency class and the site of sample collection

S.No	Name of the Organism	Cell type	Ca	Sa	Ka	Freq (%)	FC	Site
	CYANOPHYCEAE							
	CHROOCOCCACEAE							
1.	*Aphanocapsa grevelli* (Hass.)Rabenh	C	+	+	+	100	C	WL
2.	*Aphanocapsa pulchara* (Kutz.)Rabenh	C	+	+	+	100	C	ST, FR, TK
3.	*Chroococcus minor* (Kutzing)Nageli	C	-	+	-	25	R	WL
4.	*Chroococcus indicus* Bernard	C	-	+	+	50	O	ST, FR
5.	*Chroococcus minutes* (Kutzing)Nageli	C	+	+	-	50	O	TK
6.	*Chroococcus turgidus* (Kutzing)Nageli	C	+	-	-	25	R	TK
7.	*Gloeocapsa atrata* (Turp.)Kutz	U	+	+	+	100	C	ST,FR
8.	*Gloeocapsa nigrescens* Nag	U	-	+	-	25	R	TK
9.	*Microcystis flos-aquae*	C	-	-	+	25	R	FR

	(Wittr.)Kirchner							
10.	*Microcystis aeruginosa* Kutzing	C	+	+	+	100	C	TK,FR
11.	*Microcystis robusta* (Clark)Nygarrd	C	-	-	+	25	R	ST
12.	*Merismopedia punctata* Meyen	C	+	+	-	50	O	TK
13.	*Synechococcus elongates* Nag	U	+	+	+	100	C	ST,FR
14.	*Synechocystis pevalekii* Ercegovi	U	+	-	+	50	O	WL
	OSCILLATORIACEAE							
15.	*Arthrospira platensis* Gomont	F	+	-	+	50	O	TK, WL
16.	*Lyngbya kuetzinghiana* Kirchner	F	-	+	+	50	O	FR
17.	*Lyngbya subtilis* (Kutzing) Hansging	F	+	-	-	25	R	WL
18.	*Lyngbya aerugino-coerulae* Agardh ex. Gomont	F	-	+	-	25	R	FR,TK
19.	*Lyngbya confervoides* L.Ag.ex.Gomont	F	+	+	+	100	C	WL,FR
20.	*Lyngbya digueiti* Gomont	F	+	+	+	100	C	ST,WL
21.	*Lyngbya epiphytica* Hieron	F	+	+	+	100	C	FR
22.	*Lyngbya semiplena* (G.Ag.)J.Ag.ex.Gomont	F	+	+	-	50	O	WL
23.	*Lyngbya major* Menegh.ex.Gomont	F	+	+	+	100	C	ST,WL
24.	*Lyngbya majuscule* Harvey	F	-	+	+	50	O	WL

	ex. Gomont							
25.	*Lyngbya putealis* Mont. ex. Gomont	F	+	+	+	100	C	FR
26.	*Lyngbya dendrobia* Brubl.ed.Biswas	F	-	+	+	50	O	WL,TK
27.	*Oscillatoria subbrevis* Schmid	F	+	+	-	50	O	FR
28.	*Oscillatoria acuminate* Gomont	F	+	+	+	100	C	FR
29.	*Oscillatoria amoena* Gomont	F	+	+	-	50	O	FR
30.	*Oscillatoria curviceps* Agardh ex. Gomont	F	-	+	+	50	O	WL,ST
31.	*Oscillatoria limosa* Ag ex. Gomont	F	+	-	+	50	O	WL
32.	*Oscillatoria pseudogeminata* Schmid	F	+	+	+	100	C	WL,FR
33.	*Oscillatoria princeps* Vaucher ex. Gomont	F	+	+	-	50	O	WL
34.	*Oscillatoria willei* Gardner ex. Drouet	F	+	+	+	100	C	ST,FR
35.	*Phormidium retzei* Kutzing ex. Gomont	F	+	-	-	25	R	FR,TK
36.	*Phormidium ambigum* Gomont	F	-	+	+	50	O	ST,TK
37.	*Phormidium jadinianum* Gomont	F	+	+	+	100	C	FR
38.	*Phormidium fragile* (Menegh.)Gomont	F	+	+	+	100	C	WL
39.	*Phormidium mucosum* Gardner	F	-	-	+	25	R	ST
40.	*Phormidium tenue*	F	+	+	-	50	O	WL

	(Menegh.)Gomont						
NOSTOCACEAE							
41.	*Anabaena fertilissima* Prasad	F	+	+	+	100	C	WL,TK
42.	*Nostoc calcicola* ex. Born	F	+	+	+	100	C	WL
SCYTONEMATACEAE							
43.	*Plectonema nostocorum* Bornet ex. Gomont	F	+	+	-	100	C	TK
44.	*Plectonema radiosum* (Schiederm.)Gomont	F	+	+	-	100	C	TK
CHLOROPHYCEAE							
45.	*Bracteacoccus minor* (Chodat petrova)	F	+	+	-	50	O	TK
BACILLARIOPHYCEAE							
46.	*Rhizoselenia sp.* Brightwell	U	+	-	-	50	O	TK
47.	*Navicula bacillum* Ehrenberg	U	+	+	-	50	O	TK
48.	*Navicula laterostrata* Hustedt	U	+	-	-	25	R	TK
49.	*Pinnularia interrupta* Smith	U	-	+	-	25	R	TK

*SA- Sree seedeviamman temple
*CA- Cheliyandi amman temple
*KA- Arulmigu Arudra Kabaleeshwarar temple

FC- Frequency class
C- Colony

U- Unicellular
F- Filamentous

FR- Floor
ST- Statue
WL- Wall
TK- Tank

R- rare (0-25% frequency)
O- occasional (26-50 % frequency)
C- common (76-100 % frequency)

Table-3 Total number of genera and species recorded and the percentage composition of genera and species in different groups

S.No	Class	Genera		Species	
		Number	% of the total (18)	Number	% of the total (49)
1.	Cyanophyceae	14	77.77	44	89.79
2.	Chlorophyceae	1	5.55	1	2.04
3.	Bacillariophyceae	3	16.66	4	8.16

Chapter 4: Discussion

A total of 49 species, of which 44 were cyanobacteria comprising both filamentous and unicellular forms, 1 species of chlorophyceae and 4 species of bacillariophyceae were recorded. The filamentous forms were dominant when compared to the unicellular forms. This is in contrary to the work of Deepa *et al.,* 2011 who reported 42 species of cyanobacteria isolated from the different temples of Tanjavur district and reported filamentous species as dominant.

In the present study, the genera *Lyngbya, Oscillatoria* and *Phromidium* dominated the wall of the temples. This confirms the early findings of Adhikary (2000), Tripathi *et al.,* (1999) and Balu Bhavani *et al.,* (2013). Adhikary (2000) reported that different species of cyanobacteria mostly belonging to the genera *Gloeocapsa, Lyngbya, Oscillatoria* and *Tolypothrix* are the major components of the Biofilms. In the present study also the dominant genera that colonized biofilm were *Oscillatoria, Phromidium, Lyngbya, Chroococcus, Plectonema, Microcystis, Aphanocapsa* and *Gloeocapsa*. All these organisms produced copius microalgae and / or enveloped with sheath layer and occurred binding with finely textured soil particles on the wall of the temple. The microbial communities inhabiting the rock were apparently dominated by filamentous microorganisms capable of inducing carbonate formation and deposition of cements (Webb and Kamber, 2000). In the present study also the recorded species of cyanobacteria colonizing the temple wall were mostly filamentous forms like *Oscillatoria, Arthrospira, Lyngbya, Phormidium, Anaebaena, Nostoc* and *Plectonema*.

Early reports of Balu Bhavani *et al.,* (2013) and Samad and Adhikary (2008) does not reveal the presence of species of *Merismopedia* and *Arthrospira* but the present study revealed the presence of species of *Merismopedia* and *Arthrospira*.

Previous studies focused mainly on the algal flora colonizing the wall of the temple (Balu Bhavani *et al.*, 2013 and Deepa *et al.*, 2011) but the present study focused not only on the algal flora of wall but also those present in temple tank, temple floor and temple statue of the surveyed temples. This is unique to the present study. In the floor and statue of the temples, the dominant genera identified were *Lyngbya, Oscillatoria* and *Phormidium*. In the case of tank, many unicellular forms like *Aphanocapsa, Microcystis, Chroococcus, Synechococcus, Synechocystis* etc., dominated along with diatoms like *Rhizoselenia* sp., *Pinnularia* sp., and *Navicula* sp.,

The former studies by Balu Bhavani *et al.*, 2013 and Deepa *et al.*, 2011 reported only the cyanobacteria colonizing the wall of temples, but the present study is an exclusive report on the presence of *Bracteococcus* sp., of chlorophyceae and *Rhizoselenia, Pinnularia* and *Navicula* species of bacillariophyceae. Therefore the present study focused on surveying algal species colonizing temple sites regardless of their classes.

Chapter 5: Summary and conclusion

- The survey of algal flora was conducted at three different temples of Erode district namely, Sree Cheliyandiamman temple, Bhavani, Sree Seedeviamman temple, Kanchikovil and Arulmigu Arudra Kabaleeshwarar temple, Erode.
- The algal flora were collected from temple wall, temple floor and temple tank of Sree Cheliyandiamman temple and Sree Seedeviamman temple. In the case of Arulmigu Arudra Kabaleeshwarar temple, the sampled sites were temple wall, temple floor and temple statue.
- The epilithic floras were removed using soft tooth brush and razor blades. The samples were collected in sterile polythene bags and brought to laboratory for further analysis.
- The algal sample was cultured using Liquid Bold's Basal medium.
- Semi permanent slides were made for each sample in triplicates and the slides were observed in Olympus microscope. Photographs were taken using photomicrographic unit.
- A total of 49 species belonging to 18 genera were recorded and these belonged to three classes namely cyanophyceae, chlorophyceae and bacillariophyceae.
- Among the three classes cyanophyceae was was the most dominant having 14 genera (77.77%) and 44 species (89.79%). Bacillariophyceae was second highest in the order of dominance having 3 genera (16.66%) and 4 species (8.16%). Chlorophyceae revealed the least occurrence with 1 genus (5.55%) and 1 species (2.04%).
- Among the genera the dominant were *Lyngbya, Oscillatoria* and *Phormidium* which occurred in all the temples surveyed.
- Colonial, unicellular and filamentous forms were recorded. A total of 10 colonial algae, 9 unicellular algae and 30 filamentous algae were recorded in the study.
- In Sree Cheliyandiamman temple, a total of 36 species belonging to 18 genera were recorded.

- In Sree Seedeviamman temple, a total of 38 species belonging to 15 genera were recorded.
- In Arulmigu Arudra Kabaleeshwarar temple, a total of 29 species belonging to 12 genera were recorded.
- Percentage frequency calculation revealed 9 species are commonly present in all the three surveyed temples with 100% occurrence.
- Seven species are occasionally present with 50% frequency. These are encountered in two of the surveyed three temples.
- Five species are the rarely occurring species present in only one place and are with 25% frequency.
- The present study revealed the presence of species of *Merismopedia* and *Arthrospira* which were not reported in any previous studies.

Bibliography

Adhikary S.P. (2000). Epilithic cyanobacteria on the exposed rocks and walls of temples and monuments of India. *Journal microbial.* 40: 67-81.

Adhikary S.P. (2000). A preliminary survey of algae of estuaries and coastal areas in Orissa. *Seaweed Res. Utilis.* 22:1-5.

Adhikary S.P. and D.P Satapathy (1996). Tolypothrix byssoidea (Cyanophyceae / cyanobacteria) from temple rock surfaces of coastal Orissa, India. *Nova hedwigia.* 62:419-423.

Anand N. (1980). Studies on blue green algal populations of rice field. Proc.Nat.Work. Algal systems. *Indian society of Biotechnology, IIT, New Delhi.* 51-54.

Anuja J. and S. Chandra (2010). Pollution indicating algae of Thiruneermalai temple tank, Pallavaram,Chennai. *A Journal of science and technology.* 3(1):49-55.

Anuja J. and S. Chandra (2012). Studies on fresh water algae in relation to chemical constituents of Thiruneermalai temple tank near Chennai. *International journal current science.* 4:21-29.

Arino X and C. Saiz Jimenez (1996). Factors affecting the colonisation and distribution of cyanobacteria algae and lichens in ancient mortar. *Aerobiologia.* 12:9-18.

Balu Bhavani, Chockaiya Manoharan and Subramaniyan Vijayakumar (2013). Studies on diversity of cyanobacteria from temples and monuments in India. *International Journal of Environment, Ecology, Family and Urban studies.* 3(1):21-32.

Bohuslav Uher (2008). Spatial distribution of cyanobacteria and algae from the tombstone in a historic cemetery in Bratislava. *Fottea.* 9(1):81-92.

Cezar Augusto Crispim, Christina C. Gaylarde and Peitler M. Gaylarde (2004). Biofilms on church walls in the Port Alegra, RS, Brazil with special attention to cyanobacteria. *International biodeterioration and biodegradation.* 54:121-124.

Chandan Kumar Jha, Ratan Kumar, S. Venkat Kumar and V. Devi Rajeswari (2015). Extraction of natural dye from Marigold flower (*Tagetes erecta* L.) and dyeing of fabric and yarns: A focus on colorimetric analysis and fastness properties. *Der pharmacia Lettre.* 7(1):185-195.

Crispim C.A., P.M. Gaylarde and C.C. Gaylarde (2002). Algal and cyanobacterial biofilms on calcareous historic buildings. *Current Microbiology.* 46:79-82.

Crispim C.A. and C.C. Gaylarde (2005). Cyanobacteria and biodeterioration of cultural heritage: A review. *Microbial Ecology.* 49:1-9.

Deepa P., S. Jeyachandran, C. Manoharan and S. Vijayakumar (2011). Survey of epilithic cyanobacteria on the temple wall of Thanjavur district, Tamil Nadu, India. *World journal of science and technology.* 8:10-14.

Fogg G.E., W.D.P. Stewart, P. Fay and A.E. Walsby (1973). The blue green algae. *Academic press, London.* 459.

Gioia Lamenti, Piero Tiano and Luisa Tomaselli (2000). Biodeterioration of ornamental marble statues in the Boboli gardens (Florence, Italy). *Journal of applied phycology.* 12(3):427-33.

Gonsalves E.A. and V.S. Yalavigi (1959). Algae in the rhizosphere of some crop plants. *Proc. Symp. Algae. I.C.A.R., New Delhi.* 335-342.

http://en.m.wikipedia.org/wiki/Algae.

Karanda V.C, G.V.Uttekar, Priyadrshini Kamble and C.T. Karando (2012). Diversity of Cyanobacteria in biofilms on building facades of Western Maharastra. *Phycological Society*.42(2):54-58.

Lakshmi Kumari Samad and Siba Prasad Adhikary (2008). Diversity of microalgae and cyanobacteria on building facades and monuments in India. 23(2):91-114.

Lusia Tomaelli, Gioia Lamenti, Piero Tiano (2000). Biodiversity of photosynthetic microorganisms dwelling on stone monuments. *International biodeterioration and biodegradation*. 46(3):251-258.

Lynn Edward and Julia Lawler (2003). The natural paint book. *Rodale Books*. 192.

Maria Filomena Macedo, Ana Zeklia Miller, Amelia Dionisio and Cesareo Saiz Jimenez (2009). Biodiversity of cyanobacteria and green algae on monuments in the Mediterranean basin:an overview. *Microbiology*.155:3476-3490.

Miller A. and M.F. Maceda (2006). Mapping and characterization of a green biofilm inside of Vilar de Frades church (Portugal.). *International Biodeterior*. 57: 329-335.

Nitin Keshari and Siba Prasad Adhikary (2013). Characterization of cyanobacteria isolated from biofilms on stone monuments at Santiniketan, India. *The journal of bioadhesion and biofilm research*.29(5):525-536.

Nitin Keshari and Siba Prasad Adhikary (2014). Diversity of Cyanobacteria on stone monuments and building façades of India and their phylogenetic analysis. *International biodeterioration and biodegradation*. 90:45-51.

Noguerol Seoane A, A. Rifon Lastra (1997). Epilithic Phycoflora on monuments, A survey of San Esteban de Ribas de Sil Monastry (Ourense, Norwegian Spain). *Cryptogam Algol*. 18: 351-361.

Ortega Calvo J.J., M. Hernandez marine and C. Saiz Jimenez (1991). Biodeterioration of building materials by cyanobacteria and algae. *International biodeter.* 28:165-185.

Pandey V.D. (2013). Rock dwelling cyanobacteria: Survival strategies and biodeterioration of monuments. *International journal of current microbiology and applied sciences.* 2(12):519-524.

Patrizia Albertano (2012). Cyanobacterial biofilms in mouments and caves. *Ecology of cyanobacteria.* 317-343.

Ranjana Bhuvan and C.N. Saikia (2005). Isolation of color components from native dye bearing plants in North Eastern India. *Bioresource Technology.* 96(3):363-372.

Saima Umbreen, Shaukat Ali, Tanveer Hussain and Rakhshanda Nawaz (2008). Dyeing properties of natural dyes extracted from turmeric and their comparison with reactive dyeing. *Research journal of Textile and Apparel.* 12(4):1-11.

Samad L.K., S.P. Adhikary (2008). Diversity of microalgae and cyanobacteria on building facades and monuments in India. *Algae.* 23: 91-114.

Sankaran B. and E. Tiruneelagandan (2015). Microalgal diversity of Parthasarathy temple tank. *Int. J. Current Microbial. App. Science.* 4(4): 168-173.

Seenayya A.G. (1972). Ecological studies in plankton of certain fish pond of Hyderabad, India. *Hydrobiologai.* 39(1):247-271.

Sethi S.K., L.K. Jamad and S.P. Adhikary (2012). Cyanobacteria and micro algae in biological crusts on soil and sub aerial habitats of Eastern and North Eastern regions of India. *Phykos.* 42(1):1-9.

Shantanu Bhattacharya, J.K. Sahu, Binata Nayak and P.Pradhan (2011). Epilithic cyanobacteria on temples of Western Odisha. *A scientific journal of biological sciences.* 2(1-2): 47-50.

Siba Prasad Adhikary and Lubomir Kovacik (2010). Comparative analysis of cyanobacteria and microalgae in the biofilms on the exterior of stone monuments in Bratislava, Slovakia and in Bhubaneshwar, India. *Journal of Indian Botanical society*.89(1&2):19-23.

Siba Prasad Adhikary (2002). Control of epilithic cyanobacterial mats of the temples of India using algicides. *Fuer Hydrobiologie*. 143:157-171.

Sikha Mandal and Jnanendra Nath (2013). Algal colonisation and its ecophysiology on the fine sculptures of terracotta monuments of Bishnupur, West Bengal, India. *International biodeterioration and biodegradation*. 84:291-299.

Sivakumar V., D.K.N.M. Jeludine, A. Bell, D.T. Glyn and P. Mackinnon (2011). The pressure distribution along stone columns in soft clay under consolidation and foundation loading. *Geotechnique*. 61(7):613-620.

Stroem K.M. (1930). Limnological observations on Norwegian lakes, Arch. *Hydrobiol*. 21:97-124.

Subramaniyan Vijayakumar (2014). Role of cyanobacteria in biodeterioration of historical monuments- A review. *BMR microbiology*. 1(1):1-13

Tatyana A.Klochkova and Gwang Hoon Kim (2005). Ornamented resting spores of a green alga- *Chlorella* sp., collected from the stone standing Buddha statue at Junguon, Mirubsazi in korea. *Algae*.20(4).295-298.

Trikey J. and S.P. Adhikary (2005). Cyanobacteria in biological soil crusts of India. *Current Science*. 89(3):515-521.

Tripathy S.N., B.S. Tiwaari and E.R.S Talpasayi (1991). Growth of cyanobacteria (blue green algae) on urban buildings. *Energy build.* 15-16: 499-505.

Tripathy P., A. Roy, N. Anand, and S.P. Adhikary (2008). Blue green algal flora on the rock surface of temples and monuments of India. *Journal of Botanical taxonomy and geobotany*.110(1-2):133-144.

Uher B, M. Aboal and L. Kovacik (2005). Epilithic and charmo endolithic phycoflora of monuments and buildings in South Eastern Spain. *Cryptogamic Algol.* 26:275-358.

Uma Maheshwari R. and N. Anand (2003). Sensitivity of the Cyanobacterium *Tolypothrix scytonemoides* isolated from temple rock to lower water potential. *Tropical ecology.* 44(2):257-259.

Webb G.E. and B.S. Kamber (2000). Rare earth elements in Holocene recfal microbialites, a new shallow seawater proxy. *Geochimica et cosmochimica Acta.* 66: 3693-3705.

www.ingramcontent.com/pod-product-compliance
Lightning Source LLC
Chambersburg PA
CBHW030039230526
45472CB00002B/585